Fluid Power
Educational
Series

Hydraulic Motors
(In the SI Units)

Joji Parambath

Hydraulic Motors
(In the SI Units)

Copyright © 2020 Joji Parambath

All rights reserved

ISBN: 9798653842627

https://fluidsys.org

Disclaimer of Liability

The contents of this book have been checked for accuracy. Since deviations cannot be precluded entirely, we cannot guarantee full agreement. Only qualified personnel should be allowed to install and work on pneumatic and hydraulic equipment. Qualified persons are defined as persons who are authorised to commission, to ground, and tag circuits, equipment, and systems following established safety practices and standards.

Table of Contents

PREFACE

Rotary actuators are the muscle behind the rotary motions in industrial and mobile hydraulic systems. Information and data on hydraulic motors are spread in many existing textbooks as well as in documents in manufacturer's domain. A modest attempt is made here to present the information and data in one book in a simple language and structured way.

The book brings out the fundamentals and other most essential technical information related to hydraulic motors. The topics are logically arranged for a simple to the complex level progression of the subject matter. The book uses the SI system of units.

Many other fluid power topics are given in other textbooks under the fluid power educational series by the same author. A list of all the books is given at the end of this book (Page No. 59). Also, please see the details at https://jojibooks.com

Enjoy reading the book.
Your feedback is most welcome.

JOJI Parambath

Chapter 1 | Basic Hydraulic Motor Working

Hydraulic rotary actuators are positive-displacement devices that convert hydraulic energy to rotary mechanical energy. A rotary actuator, used in a hydraulic system, converts the system pressure and flow, to a controllable rotary force (torque) or rotary motion or both. Here, the fluid pressure is converted to torque, and the flow rate is converted to rotary speed. The torque or motion of the rotary actuator can be used for obtaining the rotary operation in industrial machinery.

Figure 1.1 | A graphic representation of a hydraulic motor*
Courtesy: Penton Business Media, Inc., U. S. A.
*Note: "The graphic is the copyrighted property of and is reprinted with the permission of Penton Business Media, Inc."

Hydraulic rotary actuators can be classified into two types. They are: (1) Semi-rotary actuators and (2) Motors. A semi-rotary actuator is capable of producing only limited rotation and can twist objects along a partial arc. On the other hand, a hydraulic motor is capable of producing continuous rotation and can impart continuous rotary motion to the connected load. Figure 1.1 shows the cut-section view of a hydraulic motor.

Basic Motor Operation

A hydraulic motor mainly consists of a set of moving elements, such as gears, vanes, or pistons connected to the output shaft of the motor, and enclosed in a single housing. Consider that the motor is employed in a hydraulic system. The shaft rotates when the pressurised system fluid is applied to the motor's rotating parts. In this way, the motor is capable of converting the applied pressure to rotary mechanical force and consequently driving the load attached to the motor. The fluid returns to the system reservoir, after passing through the motor. A drain connection is provided in the motor to drain the leakage fluid back to the reservoir.

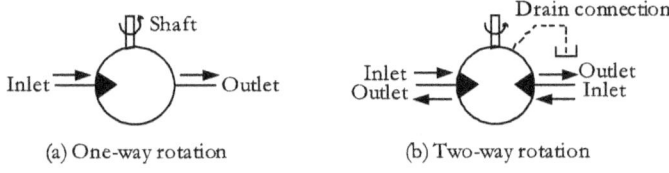

(a) One-way rotation   (b) Two-way rotation

Figure 1.2 | Schematic diagrams of hydraulic motors

Figure 1.2(a) and (b) shows the illustrative diagrams of unidirectional and bi-directional hydraulic motors, respectively. The unidirectional hydraulic motor provides rotation in only one direction, whereas the bi-directional hydraulic motor is capable of providing rotations in clockwise and anticlockwise directions. The direction of rotation of the motor's shaft can easily be reversed by changing the direction of the fluid flow through the motor ports.

This book describes various aspects of semi-rotary actuators and hydraulic motors, including their construction, classification, and operation.

Chapter 2 | Differences between Pumps and Motors

Understandably, hydraulic motors are very similar to hydraulic pumps in design and construction. However, many differences exist between the two.

The main difference between a pump and a hydraulic motor is that the moving parts in the pump, connected to a hydraulic system, push the system fluid and create flow and pressure in the system, whereas, in the motor, the pressurised fluid pushes its moving elements and produces rotary mechanical motion and force.

Another difference is that the pump is always coupled to its prime mover, whereas the motor is always coupled to a load.

Apart from that, some design modifications are required in the case of hydraulic motors, as a result of their unique requirements in applications. For example, a hydraulic motor has to overcome high starting torque at low speeds and effects of side loading.

The symbolic representations of a hydraulic pump and hydraulic motor are given in Figure 2.1.

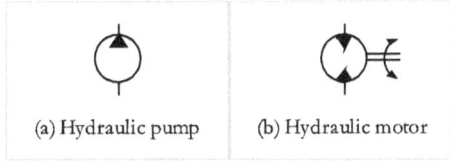

(a) Hydraulic pump   (b) Hydraulic motor

Figure 2.1 | Symbolic representations of a hydraulic pump and hydraulic motor

Chapter 3 | Terms and Definitions – Hydraulic Motors

Some critical factors relevant to the operation and applications of every hydraulic motor is its operating pressure, displacement, flow rate, input power, output power, torque output, and efficiency. The following sections describe these terms.

Operating Pressure (P): It is the pressure in a hydraulic system that overcomes all resistances in the system, which includes both useful work and losses. The rated pressure of a hydraulic motor is the maximum pressure, which the manufacturer recommends for the motor.

Motor Displacement (V_D): It refers to the volume of the system fluid required for turning the output shaft of a motor through one revolution. Some of the units of motor displacement are m^3/rev or cc/rev or in^3/rev.

Theoretical Flow Rate (Q_T): It is the quantity of the system fluid that must flow through a motor per unit of time, provided there is no leakage in the system. In the SI system of units, the flow rate is commonly measured in m^3/s or lpm. The mathematical equation for the theoretical flow rate (Q_T) of the hydraulic motor in the SI system of units is as follows:

$$Q_T (m^3/s) = V_D (m^3/rev) \times n(rps)$$

Slippage in Hydraulic Motors: It is the internal leakage of the system fluid that passes through the unintended paths of a motor, without performing any useful work. As the slippage in the hydraulic motor increases, more and more available flow intended for doing the useful work is lost, leading to the loss of power. However, all hydraulic motors are susceptible to some amount of slippage. The slippage increases, as the system pressure increases. Other reasons for the increase of slippage are the wear induced enlargement of the clearances between the internal parts of the motor and temperature-induced thinning of the system fluid.

Speed: It is directly related to the theoretical flow rate to a motor and inversely related to the displacement of the motor. It can be expressed in the SI system of units by the following equations:

$$\text{Speed, } N \text{ (rpm)} = \frac{\text{Theoretical flow to the motor, } Q_T \text{ (m}^3/\text{m)}}{\text{Motor displacement, } V_D \text{ (m}^3/\text{rev)}}$$

Therefore, it can be observed that, for a given flow rate, increasing the motor displacement decreases the motor speed and vice versa. Remember that the intended application decides the operating speed of the hydraulic motor.

Maximum motor speed is the speed of a hydraulic motor, at a particular inlet pressure, that it can sustain for a limited period without damage to the motor.

Minimum motor speed is the slowest, continuous, rotational speed obtainable from the output shaft of a hydraulic motor.

Input Power (P_{in}): Figure 3.1 gives the block diagram of a hydraulic motor, with the power relationships at its input side and output side.

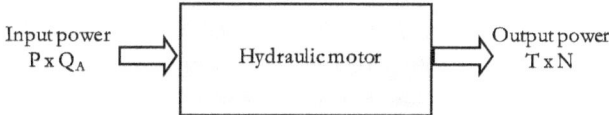

Figure 3.1 | A block diagram showing the power relationships in the hydraulic motor

The mathematical equation for the input power of the hydraulic motor, in the SI system of units, is as follows:

$$\text{Input Power, (Watt)} = P(\text{Pa}) \times Q_A(\text{m}^3/\text{s})$$

$$\text{Input Power, (kW)} = \frac{P(\text{bar}) \times Q(\text{lpm})}{600}$$

Theoretical Torque (T_T): Theoretical torque of a hydraulic motor is a function of the motor's displacement and the system pressure. The mathematical equation for the theoretical torque of the motor, in the SI system of units, is given below. The theoretical figures represent the torque available at the motor shaft, assuming no mechanical losses.

$$\text{Theoretical Torque, } T_T(\text{Nm}) = \frac{V_D(\text{m}^3/\text{rev}) \times P(\text{Pa})}{2\Pi}$$

Breakaway (Starting) Torque of a hydraulic motor is the rotary force required for turning a stationary load connected to the motor. More torque is required to turn the stationary load than that required to keep it moving. This fact is because initially, the motor has to overcome the inertia of the load. Therefore, the motor needs a breakaway (starting) torque large enough to turn the load.

Running Torque of a hydraulic motor refers to the torque required to run a load connected to the motor. Remember, the running torque of the hydraulic motor changes whenever there is a variation in the associated system pressure.

Stalling Torque of a running hydraulic motor is the torque needed to stop the motor to a standstill.

Torque ripple of a hydraulic motor is the difference between the minimum torque and maximum torque delivered by the motor at a given pressure during its one cycle of rotation.

6

Example 3.1 | A skid steer broom used to clean construction site has the hydraulic motor with a displacement of 0.0000524 m³/rev and operating at a pressure of 207 bar. What is the maximum theoretical torque the motor is capable of producing?

Solution

Volumetric displacement, V_D = 0.0000524 m³/rev
Pressure, P   = 207 bar

Theoretical Torque, T_T = V_D(m³/rev)xP(Pa)/(2$\prod$) Nm
   = 0.0000524 x 207x10⁵ / (2$\prod$)
   = 172.6 Nm

Actual Torque (T_A): It is the torque which a motor develops to drive the attached load alone. It is equal to theoretical torque minus the torque losses on account of any friction in the motor.

Output Power (P_{out}): The mathematical equation for the output power of a hydraulic motor, in the SI system of units, is as follows:

$$\text{Output Power, (Watt)} = T_A(\text{Nm}) \times \omega \text{ (rad/s)}$$

$$\text{Output Power, (kW)} = \frac{T(\text{Nm}) \times N(\text{rpm})}{9550}$$

Motor Efficiency: The efficiency of a hydraulic motor is the ratio of its output power and its input power. An ideal hydraulic motor is a motor that has no leakage and frictional losses, and it is 100% efficient. In practice, however, there are leakages and frictional losses taking place in the motor. Accordingly, two basic types of efficiencies are identified for the motor. They are (1) Volumetric efficiency, and (2) Mechanical efficiency. Overall efficiency can, then, be derived from these two types of efficiencies. The following sections briefly explain these terms.

Volumetric Efficiency (η_v) of the hydraulic motor is the ratio of the theoretical flow rate responsible for developing the actual motor speed to the total flow rate consumed by the motor, including the leakage in the motor. Remember, the motor consumes more flow than it should theoretically, due to the leakage in the motor. The mathematical equation for the volumetric efficiency of the motor is as follows:

$$\text{Volumetric efficiency, } \left(\eta_v\right) = \frac{\text{Theoretical flow rate }(Q_T)}{\text{Actual flow rate }(Q_A)}$$

Remember, the speed of a hydraulic motor is entirely dependent on the flow through the motor, and is independent of the pressure drop across the motor.

Mechanical efficiency (η_m) of the hydraulic motor is the ratio of the actual torque delivered by the motor to the theoretical torque of the motor. The hydraulic motor produces less torque than it should theoretically, due to the frictional losses in the motor. The mathematical equation for the mechanical efficiency of the hydraulic motor is as follows:

$$\text{Mechanical efficiency, } \left(\eta_m\right) = \frac{\text{Actual torque, }(T_A)}{\text{Theoretical torque }(T_T)}$$

Overall Efficiency (η_o) of the hydraulic motor is the ratio of the 'brake' power delivered by the motor to the hydraulic power delivered to the motor. Both volumetric and mechanical efficiencies reduce the overall performance of the motor. Therefore, the overall efficiency of the motor is also the product of its volumetric efficiency and its mechanical efficiency and is expressed mathematically as:

$$\text{Overall efficiency, } \left(\eta_o\right) = \frac{\text{Brake power delivered by the motor}}{\text{Hydraulic power delivered to the motor}}$$

$$= \eta_v \text{ x } \eta_m$$

8

Summary of Relations for Hydraulic Motors

Figure 3.2 gives the summary of essential relations of hydraulic motors, in the SI system units.

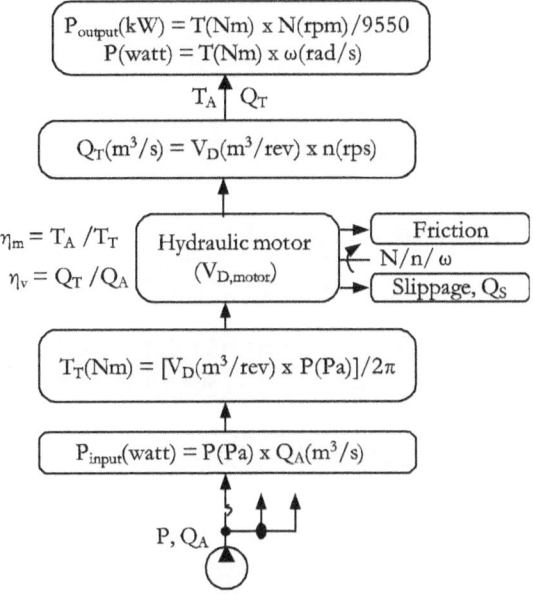

Figure 3.2 | Summary of relations for a hydraulic motor

Example 3.2 | A hydraulic gear motor consumes 44.4 lpm while running at a speed of 500 rpm. Assume the volumetric efficiency of the motor as 90%. What is the volumetric displacement of the motor?

Solution

Actual flow rate, Q_A	= 44.4 lpm = 0.00074 m³/s
Motor speed, N	= 500 rpm
Motor speed, n	= 8.33 rps
Volumetric efficiency, η_v	=0.9

Theoretical flow rate, Q_T = Q_A x η_v= 0.00074 x 0.9
                                = 0.000667 m³/s

Volumetric displacement, V_D = Q_T (m³/s) /n (rps)
                                = 0.000667 /8.33
                                = 0.00008 m³/rev

Example 3.3 | What is the actual torque supplied by a hydraulic motor of 50 cc/rev at 250 bar? Assume the mechanical efficiency of the motor as 90%.

Solution

V_D	= 50 cc/rev
	= 50x10⁻⁶ m³/rev
Pressure, P	= 250 bar
Mechanical efficiency, η_m	= 90%

Theoretical torque, T_T = V_D (m³/rev) x P (Pa) / (2Π) Nm
                          = 50x10⁻⁶ x 250x10⁵/ (2Π)
                          = 198.94 Nm

Actual torque, T_A = η_m x T_T
                                =0.9 x 198.94 =179 Nm

Example 3.4 | A hydraulic motor rotates at a speed of 450 rpm with a nominal displacement of 4 cm³/rev. The pressure differential across the hydraulic motor is 75 bar. The overall efficiency is 80%, and the volumetric efficiency is 90%. Calculate the following: (1) Theoretical flow rate, (2) Actual flow rate, (3) Power input, (4) Shaft power and (5) Shaft torque.

Solution

Speed, N	= 450 rpm
Displacement, V_D	= 4 cm³/rev
	= 4x10⁻⁶ m³/rev
ΔP	= 75 bar
η_v	= 90%
η_o	= 80%
Speed, n	= 450/60 rps = 7.5 rps
Theoretical flow rate, Q_T	= V_D (m³/rev) x n (rps)
	= 4 x10⁻⁶ x 7.5
	= 30 x10⁻⁶ m³/s
Actual flow rate, Q_A	= Q_T/ η_v
	= 30 x10⁻⁶ /0.9
	= 33.33 x10⁻⁶ m³/s
Power input P_{in}	= ΔP (Pa) x Q_A (m³/s)
	= 75 x10⁵ x 33.33 x10⁻⁶
	= 249.97 Watt
Shaft power, P_{out}	= P_{in} x η_o
	= 249.75 x 0.8
	= 199.98 Watt
Shaft torque, T	= P_{out} (watt) /2π n
	= 200 / (2πx7.5)
	= 4.24 Nm

Example 3.5 | A hydraulic motor operating at a pressure of 100 bar produces a displacement of 0.000164 m³ and a speed of 2000 rpm. If the actual flow consumed by the motor is 0.006 m³ and the motor's actual torque is 250 Nm, find its output power, volumetric efficiency, mechanical efficiency and overall efficiency.

Solution

P	= 100 bar
V_D	= 0.000164 m³
N	= 2000 rpm = 210 rad/s
Q_A	= 0.006 m³/s
T_A	= 250 Nm

Output power	= T_A (Nm) x N (rad/s)
	= 250 x 210
	= 52500 Watt

Theoretical flow rate, (Q_T)	= V_D (m³/rev) x n (rps)
	= 0.000164 x (2000/60)
	= 0.0055 m³/s

Theoretical torque, (T_T) = V_D (m³/rev) x P (Pa)/ (2Π)
= 0.000164 x 100 x 10^5/ (2Π)
= 261 Nm

Mechanical efficiency, (η_m)	= T_A/ T_T
	= 250/261.15 = 0.96

Volumetric efficiency, (η_v)	= Q_T/ Q_A
	= 0.0055/0.006 = 0.92

Overall efficiency, (η_v)	= η_v x η_m
	= 0.91 x 0.96 = 0.87

Example 3.6 | Calculate the flow rate (size) requirement of a hydraulic motor to be used in a hydraulic system for driving a load of 3730 watts at 3000 rpm. The system PRV is set at a pressure of 207 bar. Assume a return line pressure of 7 bar and mechanical efficiency of 88% and volumetric efficiency of 93% for the motor.

Solution

Output power, P_{out}	= 3730 Watt
Motor speed, N	= 3000 rpm
Motor speed, n	= N/60 = 3000/60 = 50 rps
System pressure, P_{system}	= 207 bar
Return line pressure, P_{return}	= 7 bar
Pressure differential, ΔP	= P_{system} - P_{return}
	= 200-7 = 200 bar
Volumetric efficiency, η_v	= 0.93
Mechanical efficiency, η_m	= 0.88
Overall efficiency, η_o	= η_v x η_m
	= 0.93 x 0.88 = 0.82
Input power, P_{in}	= P_{out} / η_o
	= 3730 / 0.82 = 4549 watt
Theoretical torque, T_T	= P_{in} (Watt) / $(2\pi n)$
	= 4549/ $(2\pi$ x 50)
	= 14.48 Nm
Motor displacement, V_D	= T_T x $2\prod$/ (ΔP)
	= 14.48 x $2\prod$/ $(200$x$10^5)$
	= 4.5 x 10^{-6} m^3/rev
Theoretical flow rate, Q_T	= V_D $(m^3$/rev) x n (rps)
	= 4.5 x10^{-6} x 50
	= 2.25x10^{-4} m^3/s
Actual flow rate, Q_A	= Q_T/ η_v
	=2.25 x10^{-4}/0.93
	= 2.42 x10^{-4} m^3/s

Chapter 4 | Constructional Features of Hydraulic Motors

The general constructional features of hydraulic motors are similar to that of hydraulic pumps. However, some design modifications are required in the case of hydraulic motors, as a result of their unique requirements in applications. For example, a hydraulic motor has to overcome high starting torque at low speeds and effects of side loading.

Usually, the nature of applications dictates the materials of construction of hydraulic motors. The materials of construction include cast iron, ductile iron, bronze, cast steel, and stainless steel. The unique features of hydraulic motors may include the use of rotary seals, drain connections, integrated flushing valves, and integrated brake valves.

Rotary Seals: A rotary seal is a critical element of a hydraulic motor. It is placed in the matching groove on the motor part. The housing to which the sealing is fitted and the shaft that rotates within it, constitute a sealing system. The primary purpose of the sealing system is to prevent the leakage of the system fluid through the clearances if any, between the mating parts of the motor. An ideal requirement for the sealing system is to have a perfectly concentric shaft. However, this requirement is difficult to achieve, and inevitably some eccentricity is always present. A seal is used to compensate for this eccentricity. Rotary seals are available in a wide variety of sizes, profiles, and materials.

Drain Connections: Every hydraulic motor has case drains with ports to drain the leakage fluid back to the associated system reservoir. A case drain line must be of adequate size, and it must be connected from the drain port to the reservoir directly without any restrictions. The line must be piped, must prevent siphoning and must terminate below the minimum fluid level in the reservoir.

The normal pressure of leakage fluid may not exceed 0.7 bar. Pressure surges at the case drain connection may not exceed 1.7 bar. However, a hydraulic motor may be designed deliberately to allow a small amount of internal leakage to lubricate and cool the internal parts of the motor.

Integrated Brake Valve: A hydraulic motor used in the hydrostatic transmission system of a vehicle tends to operate faster than what corresponds to the available system flow when the vehicle runs in a steep downhill. This faster operation may lead to undesirable cavitation in the motor. A brake valve can be integrated with the motor to throttle the return flow from the motor to provide the braking action of the vehicle.

Anti-cavitation Valve: A hydraulic motor is susceptible to cavitation if there is insufficient pressure at the inlet. An anti-cavitation valve, as shown in Figure 4.1, can be used to reduce the effects of cavitation. It is merely a check valve connected between the pressure and return ports. It opens to ensure flow to the motor if the inlet pressure becomes too low. It is essential to have sufficient back pressure on the return line. The motor must have a defined direction of rotation when using the anti-cavitation valve.

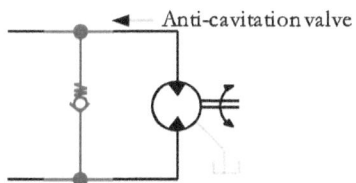

Figure 4.1 | Anti-cavitation valve connected to a motor

Integrated Flushing Valve: A hydraulic motor operating at high speeds and power levels may be provided with an integrated flushing valve to provide its rotating parts with the additional cooling flow.

Chapter 5 | Side Loads on Hydraulic Motors

The performance of hydraulic motors is affected by the presence of thrust loads and side loads. A thrust load occurs in a hydraulic motor when a compressive load acts along the longitudinal axis of the motor shaft. A side load occurs in the motor when it is coupled to the load through a pulley or gear system, or when its shaft bears the weight of the attached load. A crucial consideration during the operation of the motor is to keep the thrust load, and the side load to a minimum as these loads affect the service life of the bearings and seals used in the motor. This adverse condition can lead to the eventual shaft breakage.

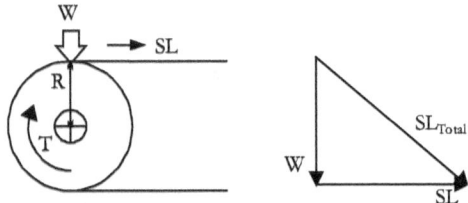

Figure 5.1 | A schematic diagram showing the side load components of a hydraulic motor

Figure 5.1 shows the sketch of the motor driving the load with a torque 'T' through a pulley of radius 'R'. Then, the side load (SL) on the motor due to the application of the load is given by:
$$SL = T/R$$

If the external load with a weight 'W' (Newton [lb]) acts on the motor shaft, then the total side load on the shaft is given by:

$$SL_{Total} = \sqrt{(W^2 + SL^2)}$$

The calculated side load on the motor at the given motor speed must be compared to the side load chart provided by the manufacturer to determine if the motor is within the permissible operating conditions or not. A motor of a larger physical size must be selected if the calculated side load goes beyond the permissible limit.

The side loads on the motor will fatigue the shaft and will place a heavy load on the shaft bearing through leverage. Bearing and shaft life can be prolonged by keeping the distance between the load line and bearing centreline as short as possible.

Example 5.1 | The motor of a Brush mower produces a torque of 120 Nm to drive pulley of 20 cm in diameter at 375 rpm. If the load with a weight of 1000 N is also acting on the motor shaft, what is the total side load due to the weight and side load on the motor?

Solution

Torque, T	$= 120$ Nm
Diameter of pulley, D	$= 20$ cm [8 in]
Weight acting on the shaft, W	$= 1000$ N

Side load, SL	$= 120 / (20 \times 10^{-2}/2)$
	$= 1200$ N

Total side load, SL_{Total}	$= \sqrt{(1000^2 + 1200^2)}$
	$= 1562$ N

Chapter 6 | Classification of Rotary Actuators

Hydraulic rotary actuators can be classified as semi-rotary actuators and motors, according to their extent of rotation. Further, they are classified according to many relevant parameters, such as the type of internal moving elements, the nature of displacement, and the torque/speed requirements, of hydraulic rotary actuators. The following sections elaborate these classifications.

Based on the type of their internal moving elements, hydraulic motors are classified as (1) Gear motors, (2) Vane motors, and (3) Piston motors.

According to the nature of displacement, hydraulic motors are divided into the fixed-displacement type or the variable-displacement type.

The fixed-displacement hydraulic motor displaces a fixed amount of system fluid with each revolution of the motor. Its displacement cannot be varied except by changing the flow rate of the system fluid. A fixed-displacement motor typically produces a constant torque. Gear, vane, and piston motors can be designed for the fixed-displacement operation. Remember, the speed of a motor is related to the displacement of the motor and the amount of fluid delivered to the motor.

The variable-displacement motor is constructed with an adjustment mechanism that can change its fluid displacement per revolution. It is possible to change the speed of the motor from zero to a maximum limit with such an adjustment mechanism, keeping the delivery of the pump a constant. The motor's torque can also be varied by varying its displacement. With the input flow and the operating pressure of the motor remaining constant, varying its displacement can change the ratio between its torque and speed to suit the load requirements. Only vane motors and piston motors can be designed and employed for the variable-displacement operation.

According to their torque-speed characteristics, hydraulic motors can be classified into two basic types.

One type is referred to as the high-speed low-torque (HSLT) motors and the other type as the low-speed high-torque (LSHT) motors.

The growing requirement of hydraulic motors with the ability to drive high-inertia loads at low speeds led to an expansion of LSHT hydraulic motor technology. A typical low-speed hydraulic motor can have its speed in the range from 0.1 to 1000 rpm and the starting torque in the range from 75% to 90% of its maximum torque. LSHT motors are known for their reliability, high-power density, and modularity.

On the other hand, a typical high-speed hydraulic motor can have its speed in the range from 1000 to 5000 rpm, but typically it does not have the high starting torque capability.

Typical high-speed motors are external gear motors, vane motors, in-line piston motors, and bent-axis piston motors.

Typical low-speed high-torque motors include gerotor motors, geroler motors, and radial piston motors.

Chapter 7 | Semi-rotary Hydraulic Actuators

A semi-rotary hydraulic actuator is a device that rotates its shaft using the fluid under pressure through a fixed arc, usually, less than 360^0. The actuator's output torque varies directly with the fluid pressure applied to it. Semi-rotary hydraulic actuators can be used to carry out various operations, such as turn, tilt, transfer, index, mix, clamp, feed, and agitate.

There are many designs of semi-rotary actuators. Figure 7.1 shows the symbol of a semi-rotary actuator, and the following sections explain the vane type, rack-and-pinion type, and helical gear-type semi-rotary actuators.

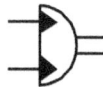

Figure 7.1 | Symbol of a semi-rotary actuator

Vane Type Semi-rotary Actuator

Figure 7.2 shows the schematic diagram of the vane type semi-rotary actuator. It consists of one or more vanes enclosed in a cylindrical chamber with a block integral to its casing for separating its high-pressure side from its low-pressure side and the necessary inlet and outlet ports. The vanes are attached to the shaft of the actuator.

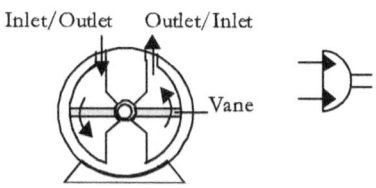

Figure 7.2 | A schematic diagram showing the cross-sectional view of a double-vane semi-rotary hydraulic actuator

When the fluid enters at one port, the vane is pushed away in one direction, and the drive shaft is turned in that direction. When the fluid enters the second port, the vane is pushed away in the opposite direction, and consequently, the drive shaft is also turned in the opposite direction.

Rack-and-Pinion Type Semi-rotary Actuator

Figure 7.3 shows the schematic diagram of the rack-and-pinion type semi-rotary hydraulic actuator. It consists of a rack-and-pinion gear assembly and a cylinder part enclosed in shared housing. The pinion gear is attached to the actuator's driveshaft.

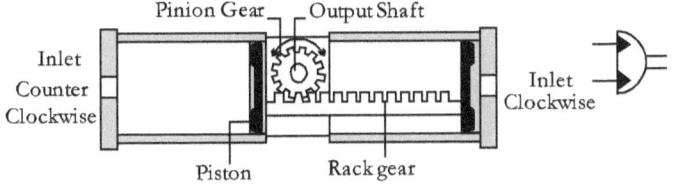

Figure 7.3 | A schematic diagram showing the cross-sectional view of a rack & pinion type semi-rotary actuator

The system fluid entering at one port of the actuator and exiting at the other port pushes the rack across the pinion gear. This action causes the rotation of the shaft in one direction through a given arc. Reversing the flows to the actuator ports reverses the direction of rotation of the shaft. The torque produced in the actuator is a function of the piston area (A), the input pressure (P), and the radius of the pinion gear (R).

The rack-and-pinion type semi-rotary actuator can turn more than one revolution as its angle of rotation is in direct relation to the pinion gear size and the rack gear length. It may also be provided with cushions and stroke limiters for its smooth adjustable stops. The rack-and-pinion type actuators are available in single-cylinder and dual-cylinder designs. The rotary motion of these actuators can be used in applications, such as turning, tilting, indexing, mixing, and clamping.

Helical Gear Type Semi-rotary Actuator

Figure 7.4 shows the schematic diagram of the helical gear semi-rotary hydraulic actuator. In this type, a helical gear is machined in the drive shaft that meshes with a matching helical gear of the non-rotating piston.

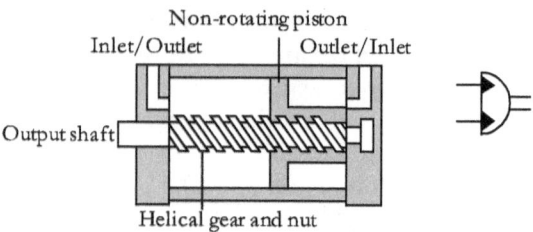

Figure 7.4 | A schematic diagram of a helical gear-type semi-rotary actuator

The fluid entering through one port imparts the output shaft a twisting action in one direction. In the same way, the fluid entering through the other port imparts the output shaft a twisting action in the opposite direction. The angle of rotation is in direct relation to the gear teeth angle and the non-rotating piston stroke. This type of semi-rotary actuator can rotate more than one revolution of its output shaft.

Chapter 8 | Hydraulic Motors

A hydraulic motor is a positive-displacement device that turns its rotating elements along with its shaft continuously using the pressurised fluid in the associated system. The motor shaft can be connected to the load either directly or through clutches and gears. The motor can develop an output torque by allowing the system pressure to act on its moving parts.

The fluid entering through the inlet port pushes the moving parts. As a result, the shaft rotates and develops a torque. The amount of torque developed by the motor depends on the system pressure, the area of the moving parts exposed to the pressurised fluid, and the distance of the moving elements from the centre of the shaft. The internal leakage of the motor is sent back to the system reservoir through a case drain.

In general, the speed of a hydraulic motor is limited by its displacement and port size. The speed of a constant displacement hydraulic motor remains constant for the constant flow of the input fluid to the motor. Whereas, the shaft torque of the motor remains constant as long as the differential pressure across the motor remains constant. Therefore, the motor's speed can be controlled by varying the flow rate, and the shaft torque can be varied by regulating the fluid pressure.

Many motors are designed for low-speed, high-torque (LSHT) operation, and some are designed for high-speed, low-torque (HSLT) operation. LSHT motors are capable of transmitting high torque with their relatively small envelopes.

Hydraulic motors have many advantages. They run smoothly at low speeds. They can be stalled or used for rapid oscillatory motion without any damage to them. Moreover, they have high energy efficiency and operate at low noise levels.

Gear Motors

Gear motors are positive-displacement devices that produce the torque by allowing the fluid pressure to act on their gears. By design, they are always fixed-displacement motors. The constructional features of the gear motors are similar to that of the gear pumps, but the operation of the gear motors is just the opposite as that of the gear pumps. That is; every gear pump pushes the system fluid to create pressure, whereas, every gear motor is pushed by the pressurised fluid to develop torque. There are two types of gear motors that have extensive applications in various machines. They are (1) external-gear motors and (2) internal-gear motors.

Gear motors have many advantages and some disadvantages. In general, they are the least expensive of all types of hydraulic motors. They can tolerate the presence of contamination in their fluid media in a better way as compared to other types of hydraulic motors. However, their efficiencies tend to be lower at lower speeds.

External-gear Motor

Figure 8.1 shows the schematic diagram of an external gear motor. It consists of a pair of matched gears enclosed in a housing, an inlet port, an outlet port, and a drive shaft. One gear unit is connected to the motor's output shaft, and the other one is an idler.

System fluid enters the housing through the inlet port. The fluid takes the path around the periphery of the housing and forces the gears to rotate. Finally, it exits through the outlet port at a low-pressure. The rotary force developed by the motor is available through its output shaft for doing some useful work. The tight-fitting gears help control the internal fluid leakage in the motor and increase the volumetric efficiency of the motor.

External gear motors are very compact and less expensive as compared to the piston and vane motors. They work best in high-speed operations and are suited for the bi-directional operation. The slippage losses in them are reasonably uniform for speeds above 500 rpm. Moreover, they can deliver relatively constant torques. However, they are the noisiest and the least efficient of the three types of hydraulic motors.

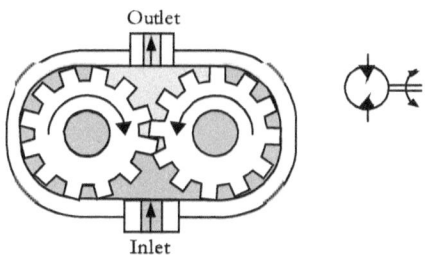

Figure 8.1 | A schematic diagram showing the cross-sectional view of an external gear motor

Gear motors are, usually, built for the high-speed, low-torque (HSLT) outputs. They are suitable for the applications with light starting loads, but, high running loads. They are very attractive as drives in many agricultural, construction, and mining equipment.

Gerotor/Geroler Motors
These types of motors are of the positive displacement type and are categorised as the internal gear type motors. They find extensive applications in mobile hydraulics.

Gerotor Motor
This motor arrangement, as shown in Figure 8.2, can be considered as a gear-with-in-a-gear type motor. It consists of two sets of interlocking gears – one set on the inner rotor and another set on an oblong rotor ring, with inlet and outlet ports, and an output shaft. The inner rotor has one tooth less than that of the rotor ring. The inner rotor is mounted on the drive shaft of the motor and is eccentric to the rotor ring. Each tooth of the rotor gear is in contact with the internal surface of the rotor ring

at all times. Fluid chambers are formed between the gear teeth and the housing.

Figure 8.2 | Schematic diagrams showing the cross-sectional views of internal gear motors

A typical gerotor motor has a 7-tooth outer ring and a 6-tooth rotor, thus forming six fluid chambers. Consider that this motor is used in a hydraulic system. At any point in time, three consecutive fluid chambers at one end are pressurised, and the three consecutive chambers at the other end are connected to the system reservoir. The high-pressure fluid is delivered to the motor through the inlet port, where it flows through the fluid chambers. As the fluid passes through the fluid chambers, both gears rotate. The drive coupling of the motor transmits the motion of the rotor to its output shaft. The fluid finally returns to the reservoir through the outlet port.

Gerotor motors are capable of providing considerable output power over a wide range of speeds. They are the most common type of low-speed, high-torque (LSHT) hydraulic motors used in industrial and mobile applications. They are used in agricultural and forestry equipment, construction machines, food processing machines, machine tools, lawnmowers, road rollers, excavators, and winches. Compact Gerotor motors are the natural choice for many applications including plastic injection moulding, CNC tool changer drives, conveyor drives, clamping, and drilling.

Geroler Motor

A variant of the gerotor motor is the Geroler motor. Figure 8.3 shows the schematic diagram of the motor. The operation of the Geroler motor is similar to that of the gerotor motor. However, the motor has its teeth fitted with rollers for reducing friction. This type of motors typically provides higher torques at lower speeds. The typical characteristic features of Geroler motors include their compact, but heavy-duty construction, smooth running even in the low-speed range, and reduced pressure spikes. These features give them a better efficiency even under high pressures and harsh operating conditions, and long service life.

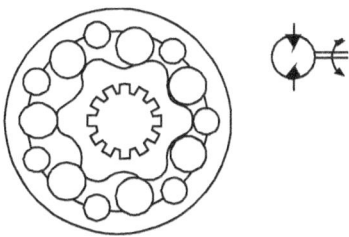

Figure 8.3 | Geroler motor

Geroler motors are well-suited for applications with the requirements of quick start and stop cycles and rapid speed reversals. An application involving the speed of less than 100 rpm should consider using a Geroler motor. The typical applications of this type of motors are found in agriculture and forestry equipment, construction machines, material handling, lifting gear and winches, machine tools, and in many other areas of industrial and mobile hydraulics.

Note: Appendix 1 gives the performance data for Geroler motors.

Vane Motor

A vane-type hydraulic motor is essentially a positive-displacement device. Figure 8.4 shows the schematic diagram of the vane motor. The constructional features of a vane motor are very similar to that of a vane pump. However, the operation of the vane motor is just the reverse as that of the vane pump. That is; the vane pump produces hydraulic power in response to the rotary mechanical power at its drive shaft, whereas the vane motor produces rotary mechanical power at its drive shaft in response to the applied hydraulic power.

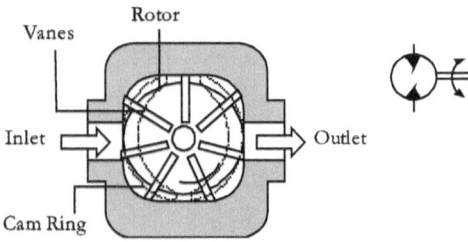

Figure 8.4 | A schematic diagram showing the cross-sectional view of a vane motor

The vane motor consists of a slotted rotor with close-fitting vanes placed in the slots. The rotor is mounted on the motor's driveshaft. It can move within the elliptical cam ring in the motor. Springs or centrifugal forces push the vanes against the cam ring. The motor also consists of two ports, one of which serves as the inlet port and the other one acts as the outlet port, according to the direction of motion required. While the fluid pushes half of the consecutive vanes, the other half helps the motor in discharging the spent fluid, through its outlet port. The motor develops the output torque by allowing fluid pressure to act on its vanes.

Vane motors work well at speeds as low as 20 rpm and as high as 6000 rpm. Moreover, they are suited for the bidirectional operation. They can operate at significantly lower noise levels as compared to other types of hydraulic motors. The cost of vane motors is higher than that of the gear motors but less than that of the piston motors, of comparable power ratings.

However, the vane motors are affected by the higher rate of fluid contamination. They are susceptible to the problem of higher internal leakages, especially at low speeds. They are also less efficient than the piston motors. The service life of the vane motors is shorter than that of the piston motors but higher than that of the gear motors.

Vane motors are extensively used on machine tools, plastic moulding machines, winches, and hydrostatic transmissions. However, they are rarely used in mobile applications outside of the high-speed and low-speed drilling applications.

Piston Motors

Piston motors are also positive-displacement devices. A piston motor consists of a cylinder block with many pistons, a cam/swash plate, and a drive shaft. In general, the designs of the piston motors are very similar to that of the piston pumps. However, the operation of the piston motors is just the reverse as that of the piston pumps. That is; every piston pump produces hydraulic power in response to the rotary mechanical power at its drive shaft, whereas every piston motor produces rotary mechanical power at its drive shaft in response to the applied hydraulic power.

Piston motors are classified according to various parameters. According to the arrangement of the cylinder blocks with respect to their drive shafts, they are classified as axial piston motors and radial piston motors. In the axial piston motor, the cylinder blocks are arranged axially, whereas, in the radial piston motor, the cylinders are arranged radially. Further, based on the motor displacement, the piston motors can be classified as fixed-displacement motors and variable-displacement motors.

Piston motors have a wide range of speeds, and they can operate typically at speeds in the range from 10 rpm to 5000 rpm with a stable torque output. They give the highest torques, speeds, and powers for the medium-duty and heavy-duty applications. They are, in general, the most efficient and versatile hydraulic motors, but they are the most expensive, as compared to other types of hydraulic motors.

Axial Piston Motors
An axial piston motor uses an axially-mounted piston block to generate mechanical power. When the high-pressure system fluid flows into the motor, the pistons are forced to move in their chambers. This action causes the motor to generate an output torque. Axial piston motors can be classified into in-line piston motors and bent-axis piston motors. Further, they can also be either the fixed-displacement type or the variable-displacement type. The fixed-displacement axial piston motor has a stationary cam plate, whereas the variable-displacement unit has some mechanical means for varying the angle of its cam plate. The torque and speed of the motor depending on the cam plate angle.

In general, the axial piston motors have excellent high-speed and speed-reversal capabilities. Moreover, they can accelerate quickly. They can provide excellent starting torque and a smooth low-speed high-torque operation. They also have excellent volumetric efficiencies, especially at lower pressures.

In-line Axial Piston Motor
In-line piston motors are the most commonly used rotary actuators in hydraulic systems. It mainly consists of a cylinder block with many pistons, an angled cam/swash plate, inlet and outlet ports, and a drive shaft. The cylinders, as shown in Figure 8.5, are arranged in a circle, parallel to each other. That is; the motor shaft and the cylinder block are aligned along the same axis. The swash plate is located at one end of the cylinder block and is acted upon by the cylinder pistons.

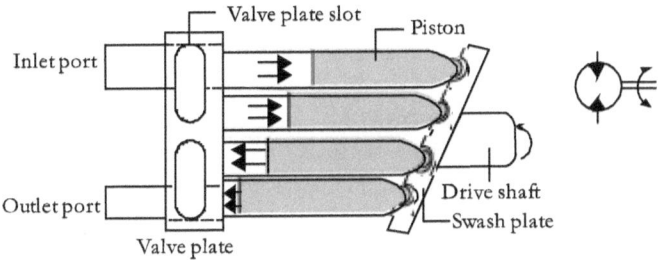

Figure 8.5 | A schematic diagram showing the cross-sectional
view of an in-line axial piston motor

The application of high-pressure system fluid at the inlet port of
the motor exerts pressure on the ends of cylinder pistons, which
then reciprocates in the cylinder block. The cylinders are filled
with the fluid in a particular sequence. This action makes the
pistons move outwards to push sequentially against the angled
swash plate and causes the cylinder block and the driveshaft to
rotate. The fluid is then swept back to the system reservoir at low
pressure on the return stroke of the piston. The torque produced
by the in-line axial piston motor is related to the swash plate
angle and the area of the pistons.

The in-line axial piston motors are built in the fixed-displacement
and the variable-displacement variants. In the fixed-displacement
type, the angle of the swash plate is set, whereas, in the variable-
displacement type, the angle of the swash plate can be varied by
various means, ranging from a simple lever to the sophisticated
servo controls. Increasing the swash plate angle improves the
torque capacity of the motor but reduces the shaft speed and
vice versa. Most manufacturers recommend a minimum swash
plate angle of 15 to 17° for best results. A maximum angle of 40
to 45° gives good torque output and extended motor life.

Axial-piston motors are well-known for their high volumetric
efficiencies during their high-speed and low-speed operations.
They are the most efficient hydraulic motors, as compared to
other types of hydraulic motors. They work best in the low-speed

high-torque (LSHT) applications. In-line axial piston motors find applications in agricultural and construction equipment.

Bent-axis Axial Piston Motor

Figure 8.6 gives the cross-sectional view of a bent-axis piston motor. It consists of a cylinder block with pistons, a cam/swash plate, inlet and outlet ports, and a drive shaft. Unlike the case of the in-line axial piston motor design, the axis of the cylinder block and the axis of the drive shaft in the bent-axis axial piston motor design are arranged at an angle to each other. The torque is developed in the motor as a reaction to the system pressure exerted on its reciprocating pistons. Because the axis of the shaft and the axis of the cylinder block are set at an angle, the force acting on the joint is resolved into axial and tangential components. The axial load is taken up by the bearings while the tangential component develops the torque at the motor's shaft. The bent-axis axial piston motor has the same operating characteristics as that of the in-line axial piston motor.

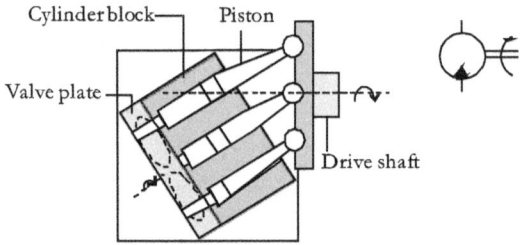

Figure 8.6 | A bent-axis axial piston motor

Bent-axis piston motors are available as the fixed-displacement and the variable-displacement types. The variable-displacement types can be controlled mechanically or by pressure compensation. The angle of the cylinder block with the drive shaft of a bent-axis piston motor determines its torque and speed ranges. The greater the angle, the higher is the torque, and the lower is the speed of the motor.

Bent-axis piston motors are rugged and capable of handling higher operating pressures. Since there is no sliding action of the piston shoes in a bent-axis piston motor, it tends to generate higher torque and less friction, for specified energy input. However, the bent-axis piston motors are hefty, particularly the variable-displacement type. They find many applications in earthmoving machines, construction equipment, forestry equipment, marine equipment, offshore equipment, industrial conveying systems, heavy-duty winches, and high-power crushers.

Note: Appendix 2 gives the performance data for Bent-axis piston motors.

Radial Piston Motors

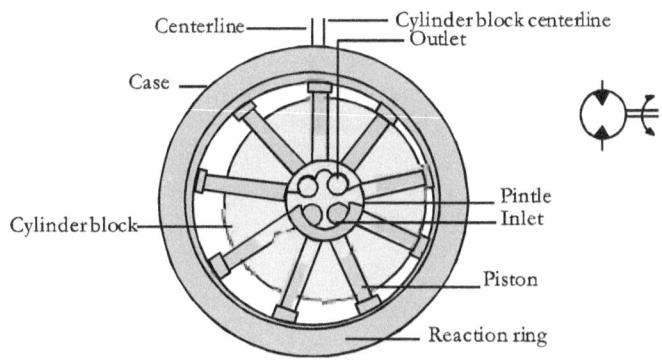

Figure 8.7 | A radial piston motor

Radial piston motors have a wide variety of designs within the primary radial configuration. Figure 8.7 shows the cross-sectional view of one type of radial piston motor for a hydraulic system. It consists of a radially arranged cylinder block with pistons arranged within the housing, a thrust ring, a pintle, and a driveshaft.

The cylinder block with five to seven radial bores is attached to the driveshaft, and the piston in each bore reciprocates when applied with pressurised fluid. The outer piston ends bear against the thrust ring.

The flow of the pressurised fluid through the pintle at the central location of the motor pushes the pistons outward. As a result, the pistons are pushed against the thrust ring, and the barrel is rotated by the reaction forces developed in the motor.

Motor displacement can be varied by shifting the cylinder block laterally. There is no fluid flow and, therefore, no rotation of the barrel, when the centerlines of the cylinder block and the housing coincide. The rotational movement of the motor can be reversed by moving the cylinder block from the existing side to the opposite side past the centre of the housing.

Radial piston motors are robust, compact, and very efficient. They have excellent low-speed capabilities and long service life. They are well-suited to start under loads. However, they have only limited high-speed capabilities and are costly. LSHT radial piston motors are used in drive applications in compact machines, such as skid steers and mini excavators. They are used as wheel motors and for other suitable applications like forklifts.

Mounting of Hydraulic Motors

Side loads on a hydraulic motor can cause excessive wear of its parts. Therefore, the motor should be mounted in such a way as to avoid side loads on its output shaft. When the side loads on the motor are unavoidable, it is necessary to support the output shaft of the motor and the attached load with auxiliary bearings. The coupling flange must be correctly aligned to the motor shaft. A flexible shaft coupling can be used, whenever possible, to avoid the side load due to the shaft misalignment.

Advantages and Disadvantages of Hydraulic Motors

Hydraulic motors have relatively large power-to-weight ratios. That means; they assist in developing compact systems. They are also simple and reliable. Each one of them can provide infinitely-variable speed controls and stalling capability even under its full load operation. The motors are also capable of reversing their rotation quickly. The main issues with them while in service are the seal failures, excessive leakage, and noise associated with them.

Selection of Hydraulic Motors

Selection of a hydraulic motor that is correctly sized for a given application depends on many factors which include system parameters and motor parameters. For example, speed and torque requirements are two essential parameters to be considered for the selection of hydraulic motors. The motor must be sized by considering the system pressure and flow rate. A hydraulic motor is often selected for its ability to operate at variable speeds merely by controlling the fluid flow to the motor.

The parameters for the selection of hydraulic motors are the system parameters, such as operating pressure range, flow, fluid viscosity, operating temperature, and noise, and the motor parameters, such as its size, weight, displacement, speed, torque, volumetric and mechanical efficiencies, mounting requirements, cost, and estimated service life.

Chapter 9 | Comparison of Hydraulic Motors

Table 9.1 gives the comparison of hydraulic motors against typical parameters in a generalised way.

Table 9.1: Comparison of hydraulic motors

Characteristics	Gear motors	Vane motors	Piston motors
General	Simple and rigid design	Most general-purpose motors	Versatile
Construction	Robust, rugged	More complicated than gear types	With tight tolerance
Size	Very compact	Compact	Heavy, bulky
Pressure	Suited for low-pressure applications	Suited for medium pressures	Suited for high-pressure applications
Speed	500-3000 rpm	100 –4000 rpm	Axial: 10-4500 rpm Radial: 0.1–2000 rpm
Low-speed operation	Not well suited	Inefficient at low speeds	Excellent choice (Radial piston motors)
High-speed operation	Best suited	Right choice	Excellent choice (Axial piston motors)
Duty	Light	Light/Medium	Heavy
Efficiency	Least efficient, at low-speed ranges	Not so efficient at low speeds	Most efficient
Cost	Least expensive	Moderate	Most expensive
Slippage	Uniform slippage losses	Higher internal leakage	Provide the best sealing
Noise	Most noisy	Least noisy	Noisy
Contamination tolerance	More tolerant	Less tolerant of contamination	Sensitive to contamination
Service life	Long	Shorter	Long

Chapter 10 | Performance Characteristics of Hydraulic Motors

The performance of a hydraulic motor is influenced by various parameters of the associated system as well as by various motor parameters. The system parameters include maximum pressure (continuous and intermittent), maximum fluid flow, maximum fluid viscosity, and operating temperature. The motor parameters include its displacement, speed (maximum and minimum), operating torque, power, volumetric and mechanical efficiencies, weight, and service life. The performance of a hydraulic motor is also affected by its working condition, duty cycle, leakage, and method of coupling. The performance can be analysed by using different characteristics, such as the torque Vs speed characteristic, the pressure Vs volumetric efficiency curve, and the flow Vs speed characteristic. The following sections give these characteristics.

Torque-Speed Characteristic

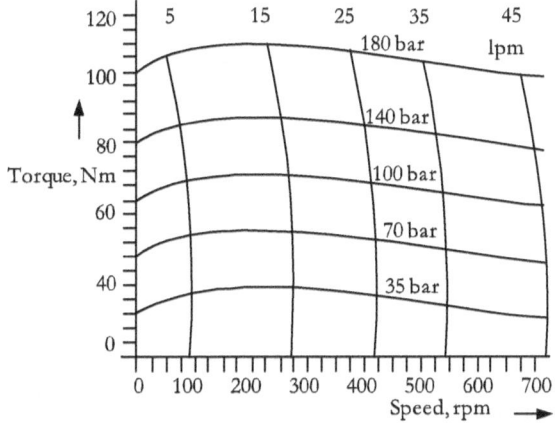

Figure 10.1 | Typical torque-speed characteristics

Figure 10.1 shows the typical torque-speed characteristics of a hydraulic piston motor. The Speed of the motor is shown on the x-axis and its torque on the y-axis. The horizontal curves correspond to the characteristics at different pressures. Sloping vertical curves represent the flow rates of the fluid through the motor.

Pressure-Volumetric Efficiency Curves
Figure 10.2 gives the pressure–volumetric efficiency curves for the gear, vane, and piston motors, for a direct comparison. For different types of motors connected in hydraulic systems, fluid leakages through them increase, and consequently, their efficiencies decrease in varying degrees, as the corresponding system pressure increases. The volumetric efficiency of the gear motor decreases linearly as the pressure increases. The volumetric efficiency of the vane motor is typically higher than that for the gear motor. Typically, the vane motor has the highest volumetric efficiency, for p < 70 bar. At higher pressures, the piston motor has higher volumetric and overall efficiencies.

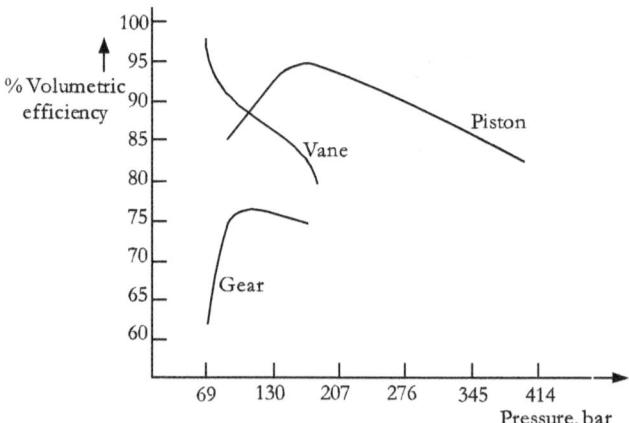

Figure 10.2 | Typical pressure-volumetric efficiency curves

Torque and Flow Curves against Speed

Figure 10.3 shows the typical flow/torque/power characteristics of a hydraulic motor plotted against its speed. The lower line is the flow Vs speed characteristic, which shows a linear relationship between the flow rate and the speed. It may be noted that the slope of the line is the displacement of the motor. The upper line is the torque Vs speed characteristic. The motor torque tends to drop off at higher speeds, as shown. Also shown in the figure is the curve for the output power of the motor against its speed.

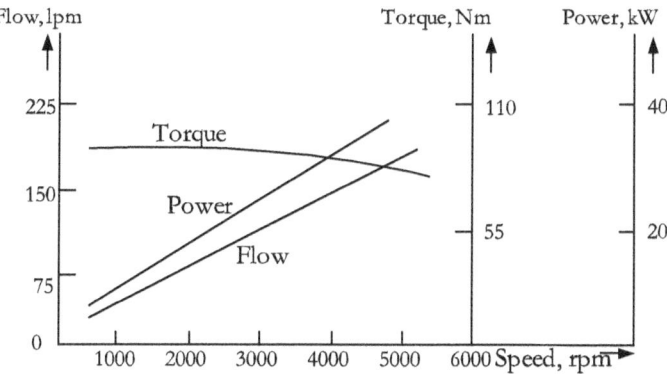

Figure 10.3 | Typical torque-speed and flow Vs speed curves

Chapter 11 | Applications of Hydraulic Motors

Hydraulic motors are most suitable for a wide range of applications due to their many positive characteristics. They are widely used for carrying out various operations, such as opening and closing, drilling, mixing, agitating, dumping, feeding, material handling, pushing, pulling and lowering, and for propelling a wheeled or track-driven vehicle and for propelling a wheeled or track-driven vehicle. They are used for heavy-duty applications including steel mills, machine tools, agriculture, construction, crane drives, winches, defence, aerospace, marine, and mining. They are also used in military vehicles, excavators, and drilling rigs. Hydraulic motors in fluid drive transmissions are becoming more popular. Figure 11.1 shows some applications of motors.

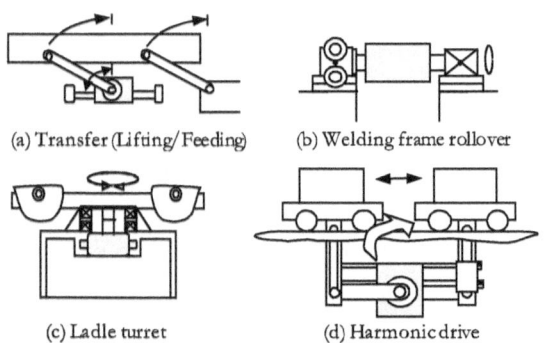

(a) Transfer (Lifting/Feeding)   (b) Welding frame rollover

(c) Ladle turret   (d) Harmonic drive

Figure 11.1 | Typical applications of hydraulic motors

The following texts highlight some industry-specific applications of hydraulic motors. In the steel industry, they tilt electric furnaces. In the construction equipment, they position tools and implements. In material handling applications, they drive conveyors. In aerospace applications, they are used to operate edge flaps and landing gears in aircraft. Next, in marine systems, they open and close hatches, and operate booms. In the underground mining machinery, they can be used to place percussion drill.

Chapter 12 | Maintenance of Hydraulic Motors

Most of the maintenance considerations of hydraulic motors are same as that for the hydraulic pumps. The most important reasons for the premature failure of hydraulic motors are dirt, corrosion, misalignment, loose bolting, and overload. It is also necessary to check the drive shaft of the motor periodically for any misalignment or damage. See Table 12.1.

Table 12.1 | Troubleshooting chart of hydraulic motors

Disturbances	Possible causes	Rectification
Motor rotating in the wrong direction	Incorrect piping between the control valve and the motor	-Connect the circuit as per the circuit diagram
Motor not developing proper speed or torque	Incorrect setting of PRV	-Set PRV correctly
	PRV sticking/open	-Remove dirt under poppet
	Free circulation of fluid to the reservoir	-Repair or replace DC valve
	Pump not delivering sufficient fluid	-Repair pump
Fluid contamination	Filters clogged	-Replace filter elements -Clean strainer -Carry out fluid analysis
External fluid leakage	Worn seals	-Replace worn seals
Fluid getting heated up	Inefficiency of system	-Check the heat exchanger
Misalignment	Mounting loosened	-Tighten the mountings -Re-check the alignment

13 | Objective Type Questions

1. Which of the following statements is <u>incorrect</u>?
   a) Gear motors work best in high-speed low-torque applications.
   b) Gear motors have higher volumetric efficiency as compared to other types of motors.
   c) Hydraulic vane motors operate at a lower noise level than other types of motors.
   d) Piston motors are the most efficient, but most expensive motors.

2. Which of the following hydraulic motors is more tolerant of contamination?
   a) Gear motor
   b) Vane motor
   c) Axial piston motor
   d) Radial piston motor

3. Which of the following hydraulic motors is an excellent choice for high-speed operation?
   a) Gear motor
   b) Vane motor
   c) Axial piston motor
   d) Radial piston motor

4. Identify the following component:

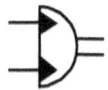

Figure 13.1

   a) Semi-rotary pneumatic actuator
   b) Semi-rotary hydraulic actuator
   c) Hydraulic motor
   d) Pneumatic motor

5. Which power transmission system does provide step-less control of speed, torque, and power?
   a) Mechanical
   b) Electrical
   c) Hydrostatic
   d) Pneumatic

6. Identify the following component:

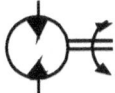

Figure 13.2

   a) Semi-rotary pneumatic actuator
   b) Semi-rotary hydraulic actuator
   c) Hydraulic motor
   d) Pneumatic motor

14 | Review Questions

1. Define hydraulic motors?
2. What are the main reasons for using hydraulic rotary actuators in industrial and mobile hydraulic systems?
3. Explain the working principles of a hydraulic motor.
4. Draw the symbols for the hydraulic semi-rotary actuators and motors.
5. Write two differences between the hydraulic motors and the hydraulic pumps.
6. Define the term 'operating pressure' of a hydraulic motor.
7. What determines the selection of a hydraulic motor operating speed?
8. Define hydraulic motor torque.
9. Define the following terms of hydraulic motors: (a) starting torque, (b) running torque, and (c) stalling torque.
10. What factors determine the torque output of a hydraulic motor?
11. Write the formula for calculating the torque of a hydraulic motor.
12. What determines the operating speed of a hydraulic motor?
13. Define the fluid power and the shaft power of a hydraulic motor.
14. Write the formula for calculating the input horsepower in the SI system units?
15. Write the formula to calculate the mechanical horsepower output of a hydraulic motor in the SI system units?
16. Define the nominal displacement of hydraulic motors.
17. Explain the relationship between the flow rate and the shaft speed of a hydraulic motor.
18. What is an ideal hydraulic motor?
19. Define the following terms of hydraulic motors: (1) Motor displacement, (2) Volumetric efficiency, and (3) Mechanical efficiency.
20. Define and explain the terms concerning hydraulic motors: (1) Volumetric efficiency, and (2) overall efficiency.
21. Write the formula for calculating the overall efficiency of a hydraulic motor.

22. Explain the interaction of the flow rate and the pressure during the operation of a hydraulic motor.
23. Determine the power, torque, and flow rates of hydraulic motors.
24. What are the functions of hydraulic rotary seals?
25. Give a brief note on the case drain connection in hydraulic motors.
26. List two differences between the hydraulic pumps and the hydraulic motors
27. How are hydraulic motors classified?
28. What are the different ways of constructing hydraulic motors?
29. How can the output speed of a variable-displacement motor be changed?
30. Name two types of positive-displacement hydraulic motors.
31. Describe the general constructional features of one type of semi-rotary hydraulic actuators.
32. Explain the working of a vane type semi-rotary hydraulic actuator with a simple sketch.
33. Explain the working of a rack-and-pinion type semi-rotary hydraulic actuator with a sketch.
34. Explain the operation of a gear motor.
35. Describe the general constructional features of gear motors.
36. Give two advantages and disadvantages of gear motors.
37. What are the applications of gear motors?
38. Explain the operation of a gerotor motor.
39. Describe the general constructional features of gerotor motors.
40. What are the advantages and disadvantages of gerotor motors?
41. What are the applications of gerotor motors?
42. Explain the construction features of a Geroler motor.
43. What are the typical characteristics of Geroler motors?
44. Describe the operation of a hydraulic vane motor.
45. Describe the general constructional features of hydraulic vane motors.

46. Give the advantages and disadvantages of hydraulic vane motors?
47. What are the applications of hydraulic vane motors?
48. Explain the difference between the vane motor and the vane pump, as used in a hydraulic system.
49. Explain the working of an in-line axial piston motor.
50. Describe the general constructional features of in-line axial piston motors.
51. What are the advantages and disadvantages of in-line axial piston motors?
52. What are the applications of in-line motors?
53. How is the fluid displacement adjusted in a variable-displacement axial-piston motor?
54. Explain the operation of a bent-axis axial piston motor.
55. Describe the general constructional features of bent-axis axial piston motors.
56. What are the advantages and disadvantages of bent-axis axial piston motors?
57. What are the applications of bent-axis axial piston motors?
58. Explain the operation of a radial piston motor.
59. Describe the general constructional features of radial piston motors.
60. What are the advantages of radial piston motors?
61. What are the applications of radial piston motors?
62. Draw the torque-speed characteristics of a hydraulic motor and explain.
63. Draw the pressure-volumetric efficiency curves of the gear, vane and piston motors and compare.
64. Write two factors that affect the rating and selection of a hydraulic motor for an application?
65. Briefly explain the areas applications of hydraulic motors.

Objective type questions - answer key:
1-b, 2-a, 3-c, 4bc, 5-c, 6-c

15 | Numerical Problems

1. A hydraulic motor theoretically consumes 13.33 cc/s while running at a speed of 2000 rpm. What is the volumetric displacement of the motor? [Ans: 0.4 cm³/rev]

2. A hydraulic piston motor operates with a pressure drop of 276 bar cross its ports. Measured flow rate to the motor is 0.003333 m³/s. What is the input hydraulic power?
[Ans: 92 kW]

3. A hydraulic piston motor must produce an output power of 78 kW with a pressure differential of 200 bar. The overall efficiency is 91%. What is the flow rate of fluid required?
[Ans: 4.285 dm³/s]

4. What is the output power of a hydraulic vane motor, if the measured torque produced by the motor is 60 Nm while running at 1000 rpm? [Ans: 6848 Watt]

5. Calculate the output power of hydraulic gear motor that has an actual flow rate of 5 dm³/s and a pressure differential of 80 bar. The overall efficiency is 65%.
[Ans: 26 kW]

6. A hydraulic motor operates with a pressure drop of 138 bar across its ports. Measured flow rate to the motor is 38 lpm. What is the input hydraulic power? If the measured torque is 122 Nm at 536 rpm, what is the output power? Also, calculate the overall efficiency.
[Ans: P_{in}= 8740 Watt, P_{out} =6848 Watt, η_o= 78%]

7. The torque required for turning a sugar mill drive using a hydraulic motor is 205 Nm. The drive unit is powered by a power pack delivering 210 lpm at 200 bar. Assume a mechanical efficiency of the motor as 88%. What is the required displacement and speed of the hydraulic motor?
[Ans: V_D=7.3199x 10^{-5} m³/rev, N=2870 rpm]

Appendix 1

The data is extracted from the actual catalogue listing of some prominent manufacturers.

Performance Data of Gerolor Motors for continuous duty

Table A1.1 | Values of Torque in Nm (Speed in rpm)

$V_D = 8.2$ cm³/rev							
ΔP(bar) Q(lpm)	14	28	41	55	69	103	140
3.8	1 (456)	3 (444)	5 (429)	6 (412)	8 (394)	12 (332)	15 (239)
7.6	1 (897)	3 (886)	4 (867)	6 (847)	8 (823)	12 (749)	16 (647)
11.4	1 (1349)	2 (1331)	4 (1309)	6 (1285)	7 (1261)	12 (1176)	16 (1060)
16.0		2 (1902)	3 (1873)	5 (1846)	7 (1817)	11 (1721)	15 (1585)
17.0		2 (1992)	3 (1964)	5 (1929)	7 (1900)	11 (1808)	15 (1673)

Table A1.2 | Values of Torque in Nm (Speed in rpm)

$V_D = 12.9$ cm³/rev							
ΔP(bar) Q(lpm)	14	28	41	55	69	103	140
3.8	2 (290)	5 (285)	7 (277)	10 (268)	12 (260)	19 (230)	25 (189)
7.6	2 (573)	4 (566)	7 (555)	10 (544)	12 (534)	19 (490)	25 (437)
11.4	1 (859)	4 (849)	7 (838)	9 (825)	12 (810)	18 (763)	25 (701)
15.1	1 (1153)	3 (1140)	6 (1129)	9 (1117)	11 (1101)	18 (1044)	25 (975)
20.8		2 (1575)	5 (1556)	7 (1539)	10 (1521)	17 (1457)	24 (1387)

Table x A1.3 | Values of Torque in Nm (Speed in rpm)

$V_D = 19.8$ cm³/rev							
ΔP(bar) Q(lpm)	14	28	41	55	69	103	140
3.8	4 (189)	8 (187)	12 (185)	15 (182)	19 (179)	29 (169)	37 (138)
7.6	3 (379)	7 (375)	11 (370)	15 (366)	19 (361)	29 (347)	38 (309)
11.4	2 (569)	6 (565)	11 (560)	14 (556)	18 (551)	28 (523)	37 (484)
15.1	1 (761)	5 (758)	9 (751)	13 (746)	17 (741)	27 (707)	36 (656)
20.8		4 (1043)	8 (1035)	11 (1028)	15 (1021)	25 (990)	34 (934)

Table A1.4 | Values of Torque in Nm (Speed in rpm)

$V_D = 31.6$ cm³/rev							
ΔP(bar) Q(lpm)	14	28	41	55	69	103	121
3.8	6 (118)	12 (116)	18 (113)	24 (111)	30 (107)	43 (81)	50 (70)
7.6	5 (236)	12 (234)	18 (230)	24 (225)	30 (221)	44 (175)	50 (165)
11.4	4 (355)	11 (352)	17 (347)	23 (342)	29 (336)	43 (287)	50 (273)
15.1	3 (474)	9 (472)	15 (466)	21 (460)	28 (452)	41 (393)	48 (373)
20.8		6 (650)	13 (645)	19 (636)	25 (629)	38 (575)	45 (550)

Table A1.5 | Values of Torque in Nm (Speed in rpm)

$V_D = 50.0$ cm³/rev							
ΔP(bar) Q(lpm)	14	28	41	55	69	83	97
3.8	9 (75)	19 (72)					
7.6	8 (149)	18 (147)	28 (144)	37 (142)			
11.4	6 (221)	16 (220)	26 (217)	35 (213)	45 (209)	55 (201)	62 (191)
15.1	3 (296)	14 (292)	23 (286)	33 (282)	42 (273)	52 (265)	61 (259)
20.8		9 (393)	19 (389)	29 (383)	38 (377)	48 (369)	57 (358)

Table A1.6 | Values of Torque in Nm (Speed in rpm)

$V_D = 36$ cc/rev							
ΔP(bar) Q(lpm)	14	28	41	55	69	97	124
7.6	6 (204)	12 (201)	18 (198)	21 (194)	31 (189)	43 (177)	55 (162)
15.1	5 (408)	12 (407)	18 (402)	22 (399)	31 (394)	43 (381)	56 (365)
22.7	5 (613)	12 (612)	18 (609)	21 (604)	31 (599)	43 (586)	56 (565)
30.3	5 (817)	11 (817)	17 (814)	21 (807)	31 (799)	43 (785)	56 (762)
37.9	4 (1021)	10 (1021)	17 (1015)	20 (1008)	30 (1001)	43 (981)	56 (959)

Table A1.7 | Values of Torque in Nm (Speed in rpm)

$V_D = 46$ cc/rev							
ΔP(bar) Q(lpm)	14	28	41	55	69	97	124
7.6	7 (161)	15 (158)	24 (156)	32 (153)	40 (148)	56 (139)	72 (127)
15.1	7 (323)	16 (320)	24 (316)	32 (314)	41 (310)	57 (300)	73 (287)
22.7	7 (486)	15 (481)	23 (479)	32 (475)	40 (471)	57 (461)	73 (444)
30.3	6 (648)	14 (643)	23 (640)	31 (635)	40 (628)	57 (617)	74 (599)
37.9	5 (808)	13 (803)	22 (798)	30 (793)	39 (787)	56 (771)	73 (753)
45.4	4 (969)	12 (964)	21 (960)	29 (952)	38 (946)	56 (931)	73 (914)

Table A1.8 | Values of Torque in Nm (Speed in rpm)

V_D = 59 cc/rev							
ΔP(bar) Q(lpm)	14	28	41	55	69	97	124
7.6	9 (127)	19 (125)	29 (123)	34 (121)	49 (117)	70 (109)	90 (96)
15.1	9 (254)	19 (254)	29 (251)	35 (249)	50 (246)	70 (236)	90 (224)
22.7	8 (381)	18 (381)	28 (380)	34 (377)	50 (373)	70 (364)	91 (349)
30.3	7 (508)	17 (508)	27 (508)	33 (504)	48 (500)	69 (491)	90 (476)
37.9	6 (635)	16 (635)	26 (634)	32 (630)	47 (626)	68 (614)	89 (601)
45.4	5 (762)	15 (762)	26 (762)	31 (757)	46 (753)	67 (741)	88 (728)
53.0	4 (889)	13 (889)	24 (887)	30 (882)	45 (877)	66 (866)	87 (851)
56.8	3 (953)	13 (953)	23 (951)	29 (945)	44 (940)	65 (929)	86 (913)

Table A1.9 | Values of Torque in Nm (Speed in rpm)

V_D = 74 cc/rev							
ΔP(bar) Q(lpm)	14	28	41	55	69	97	124
7.6	12 (101)	25 (99)	38 (98)	51 (96)	64 (93)	91 (86)	117 (76)
15.1	11 (203)	25 (201)	38 (199)	52 (197)	65 (194)	91 (187)	118 (177)
22.7	11 (305)	24 (303)	37 (301)	51 (298)	85 (296)	91 (288)	118 (276)
30.3	10 (406)	22 (404)	36 (402)	49 (399)	63 (396)	90 (388)	117 (377)
37.9	8 (507)	21 (505)	35 (502)	48 (499)	62 (496)	89 (486)	116 (476)
45.4	7 (608)	19 (606)	33 (603)	46 (600)	60 (596)	87 (587)	114 (576)
53.0	5 (709)	17 (706)	31 (702)	45 (698)	58 (694)	86 (686)	113 (674)
56.8	4 (760)	16 (757)	30 (753)	44 (749)	57 (744)	85 (735)	113 (723)

Table A1.10 | Values of Torque in Nm (Speed in rpm)

V_D = 97 cc/rev							
ΔP(bar) Q(lpm)	14	28	41	55	69	97	124
7.6	15 (78)	33 (76)	50 (75)	67 (73)	84 (71)	119 (65)	154 (55)
15.1	15 (156)	32 (155)	49 (153)	67 (151)	85 (149)	120 (143)	154 (134)
22.7	14 (234)	30 (233)	48 (231)	66 (230)	84 (228)	119 (221)	155 (210)
30.3	12 (312)	28 (311)	46 (310)	64 (308)	81 (305)	116 (300)	151 (291)
37.9	11 (390)	26 (389)	44 (387)	61 (385)	79 (383)	114 (376)	149 (368)
45.4	9 (468)	25 (467)	43 (465)	59 (463)	77 (460)	112 (453)	147 (445)
53.0	7 (546)	22 (544)	40 (542)	58 (539)	75 (537)	110 (531)	146 (521)
56.8	6 (585)	21 (583)	39 (581)	56 (578)	74 (575)	109 (589)	145 (559)

Table A1.11 | Values of Torque in Nm (Speed in rpm)

V_D = 120 cc/rev							
ΔP(bar) Q(lpm)	14	28	41	55	69	97	124
7.6	18 (62)	40 (61)	61 (61)	83 (59)	105 (58)	147 (53)	191 (45)
15.1	18 (125)	39 (124)	61 (123)	83 (121)	105 (120)	149 (116)	192 (110)
22.7	18 (188)	38 (187)	60 (186)	82 (185)	104 (183)	148 (178)	192 (170)
30.3	16 (250)	36 (250)	58 (249)	80 (247)	102 (245)	145 (241)	189 (233)
37.9	14 (313)	34 (312)	56 (311)	78 (309)	100 (308)	143 (302)	187 (296)
45.4	12 (375)	33 (374)	54 (373)	75 (371)	97 (370)	141 (365)	185 (358)
53.0	9 (438)	30 (437)	52 (435)	74 (433)	95 (431)	139 (427)	183 (419)
56.8	8 (469)	29 (468)	50 (466)	71 (464)	94 (462)	137 (458)	182 (450)

Table A1.12 | Values of Torque in Nm (Speed in rpm)

ΔP(bar) Q(lpm)	$V_D = 146$ cc/rev						
	14	28	41	55	69	97	117
7.6	22 (51)	49 (50)	75 (50)	101 (49)	128 (47)	180 (43)	219 (39)
15.1	22 (103)	48 (102)	74 (101)	101 (99)	128 (99)	181 (95)	221 (92)
22.7	21 (154)	47 (153)	73 (152)	100 (151)	127 (150)	181 (146)	220 (141)
30.3	19 (205)	44 (205)	71 (204)	98 (203)	124 (201)	177 (197)	217 (193)
37.9	17 (257)	42 (256)	68 (255)	94 (253)	122 (252)	174 (248)	215 (244)
45.4	14 (308)	40 (307)	66 (306)	92 (305)	119 (303)	172 (299)	212 (295)
53.0	11 (359)	36 (358)	63 (357)	90 (355)	116 (354)	169 (350)	209 (346)
56.8	9 (385)	35 (384)	61 (383)	87 (381)	114 (379)	167 (375)	208 (371)

Table A1.13 | Values of Torque in Nm (Speed in rpm)

ΔP(bar) Q(lpm)	$V_D = 159$ cc/rev						
	14	28	41	55	69	97	134
7.6	24 (47)	53 (46)	81 (46)	110 (45)	139 (44)	195 (40)	231 (37)
15.1	24 (94)	52 (94)	80 (93)	110 (91)	139 (91)	197 (89)	233 (87)
22.7	23 (141)	51 (141)	80 (140)	109 (139)	137 (138)	196 (134)	232 (132)
30.3	21 (188)	50 (188)	78 (187)	107 (186)	136 (185)	194 (181)	230 (178)
37.9	19 (235)	48 (234)	76 (234)	105 (232)	134 (232)	192 (228)	230 (225)
45.4	16 (282)	46 (281)	74 (281)	102 (279)	131 (279)	189 (275)	226 (272)
53.0	12 (330)	42 (329)	70 (328)	100 (327)	129 (325)	187 (322)	222 (319)
56.8	10 (353)	41 (352)	69 (351)	97 (350)	127 (348)	185 (345)	220 (342)

Table A1.14 | Values of Torque in Nm (Speed in rpm)

$V_D = 185$ cc/rev							
ΔP(bar) Q(lpm)	14	28	41	55	69	97	110
7.6	29 (40)	63 (40)	96 (39)	130 (38)	163 (37)	230 (33)	262 (29)
15.1	29 (81)	62 (81)	95 (80)	129 (79)	164 (78)	232 (76)	265 (74)
22.7	28 (121)	61 (121)	94 (120)	128 (120)	162 (119)	230 (115)	264 (112)
30.3	25 (162)	59 (162)	93 (161)	126 (160)	160 (159)	228 (155)	262 (152)
37.9	23 (202)	56 (202)	90 (201)	124 (201)	158 (200)	226 (196)	260 (193)
45.4	20 (243)	54 (242)	87 (242)	120 (241)	155 (240)	222 (236)	256 (234)
53.0	16 (283)	50 (283)	83 (282)	117 (281)	151 (280)	219 (277)	252 (274)
56.8	14 (304)	48 (303)	81 (302)	115 (301)	149 (300)	216 (297)	249 (294)

Table A1.15 | Values of Torque in Nm (Speed in rpm)

$V_D = 231$ cc/rev							
ΔP(bar) Q(lpm)	14	28	41	55	69	97	100
7.6	38 (32)	80 (32)	121 (31)	165 (30)	206 (30)	291 (26)	300 (25)
15.1	37 (65)	79 (65)	122 (64)	163 (63)	206 (62)	291 (60)	302 (60)
22.7	36 (97)	78 (97)	119 (97)	162 (96)	205 (95)	289 (92)	299 (91)
30.3	33 (130)	74 (130)	117 (130)	159 (129)	201 (128)	286 (124)	297 (124)
37.9	30 (162)	71 (162)	113 (162)	156 (162)	198 (160)	284 (156)	294 (156)
45.4	26 (195)	68 (195)	109 (194)	152 (194)	195 (193)	280 (189)	290 (189)
53.0	22 (227)	64 (227)	105 (227)	147 (226)	190 (226)	274 (222)	285 (221)
56.8	19 (243)	61 (243)	102 (243)	145 (242)	188 (242)	272 (238)	282 (238)

Table A1.16 | Values of Torque in Nm (Speed in rpm)

ΔP(bar) Q(lpm)	$V_D = 293$ cc/rev						
	14	28	41	55	69	83	93
7.6	48 (26)	101 (25)	154 (25)	207 (24)	259 (22)	302 (16)	336 (13)
15.1	47 (51)	100 (51)	154 (51)	207 (50)	260 (49)	313 (47)	351 (44)
22.7	45 (77)	99 (77)	152 (76)	206 (76)	259 (74)	312 (71)	350 (68)
30.3	41 (102)	95 (102)	149 (102)	202 (101)	254 (100)	308 (98)	347 (95)
37.9	38 (128)	91 (128)	144 (128)	198 (127)	250 (126)	303 (123)	343 (120)
45.4	33 (153)	86 (153)	139 (153)	193 (153)	246 (151)	298 (149)	338 (146)
53.0	27 (179)	80 (179)	133 (179)	186 (179)	240 (177)	293 (175)	332 (172)
56.8	24 (192)	77 (192)	130 (192)	183 (191)	237 (190)	289 (188)	328 (185)

Table A1.17 | Values of Torque in Nm (Speed in rpm)

ΔP(bar) Q(lpm)	$V_D = 370$ cc/rev						
	14	28	41	55	69	83	86
7.6	61 (20)	127 (20)	194 (20)	258 (19)	323 (16)		
15.1	60 (40)	127 (40)	194 (40)	261 (39)	327 (38)	392 (36)	407 (35)
22.7	57 (61)	124 (61)	191 (61)	259 (60)	326 (58)	391 (55)	407 (53)
30.3	52 (81)	120 (81)	188 (81)	255 (80)	321 (79)	386 (76)	402 (74)
37.9	47 (101)	115 (101)	182 (101)	249 (101)	315 (99)	380 (96)	396 (94)
45.4	41 (121)	108 (121)	175 (121)	243 (121)	309 (119)	373 (116)	389 (115)
53.0	34 (142)	101 (142)	168 (142)	236 (142)	301 (140)	367 (137)	383 (136)
56.8	30 (152)	97 (152)	164 (152)	232 (152)	297 (150)	362 (148)	378 (147)

Table A1.18 | Values of Torque in Nm (Speed in rpm)

$V_D = 739$ cc/rev			
ΔP(bar) Q(lpm)	14	28	41
2	122 (10)	254 (10)	389 (10)
4	121 (20)	254 (20)	389 (19)
6	115 (30)	249 (30)	383 (29)
8	107 (40)	241 (40)	376 (39)
10	95 (50)	232 (50)	367 (48)
12	84 (60)	220 (59)	354 (58)
14	71 (69)	206 (68)	340 (68)
15	61 (74)	196 (74)	328 (73)

Appendix 2

Typical Specifications – Bent-axis Axial Piston motors

Table A2.1

Displacement, cc/rev	Maximum pressure, bar (Continuous)	Maximum speed, rpm	Motor continuous input flow, lpm
4.9	350	10800	41
9.8	350	9900	67
14.3	350	9000	87
19.0	350	8100	103
30.0	350	5600	168
40.0	350	5000	200
59.8	350	4300	257
80.4	350	4000	322
110.1	350	3600	396
150	350	2600	390
242	350	2400	580

16 | References

1. Andrew Parr, Hydraulics & Pneumatics, A technician's and engineer's guide, 2nd Edition, Butterworth, Heinemann, 1998

2. Anthony Esposito, Fluid Power with Applications, 6th Edition, Prentice-Hall of India, 2006

3. Article on 'About Hydraulic Motors', GlobalSpec Inc. 350 Jordan Rd, Troy, NY, USA

4. Article on 'Bigger Isn't Always Better When It Comes to Sizing Your Hydraulic Motors for Efficiency', By Phillip Groves, White Hydraulics Inc., Hopkinsville, Ky., in Compact Equipment, October 2003

5. Article on 'How Does a Hydraulic Motor Work', by Shashank Nakate in Buzzle.com

6. Article on 'Hydraulic motors – Part 1' and 'Hydraulic motors – Part 2', Penton Media, Inc. & Hydraulics & Pneumatics magazine

7. Catalogue on Low Speed, High Torque Motors (Document No. E-MOLO-MC001-E9) EATON Corporation, USA

8. Eaton Hydraulics Training Services, Mobile Hydraulic Manual, 2010

9. Publications Department of Womac Machine Supply Company, Industrial Fluid Power, Volume 3 Third Edition

Fluid Power Educational Series Books

1. Pneumatic Systems and Circuits -Basic Level (In the SI Units)
2. Industrial Pneumatics -Basic Level (In the English Units)
3. Pneumatic Systems and Circuits -Advanced Level
4. Electro-Pneumatics and Automation
5. Design of Pneumatic Systems (In the SI Units)
6. Design Concepts in Pneumatic Systems (In the English Units)
7. Maintenance, Troubleshooting, and Safety in Pneumatic Systems
8. Industrial Hydraulic Systems and Circuits -Basic Level (In the SI Units)
9. Industrial Hydraulics -Basic Level (In the English Units)
10. Hydraulic Fluids
11. Hydraulic Filters: Construction, Installation Locations, and Specifications
12. Hydraulic Power Packs (In the SI Units)
13. Power Packs in Hydraulic Systems (In the English Units)
14. Hydraulic Cylinders (In the SI Units)
15. Hydraulic Linear Actuators (In the English Units)
16. Hydraulic Motors (In the SI Units)
17. Hydraulic Rotary Actuators (In the English Units)
18. Hydraulic Accumulators and Circuits (In the SI Units)
19. Accumulators in Hydraulic Systems (In the English Units)
20. Hydraulic Pipes, Tubes, and Hoses (In the SI Units)
21. Pipes, Tubes, and Hoses in Hydraulic Systems (In the English Units)
22. Design of Industrial Hydraulic Systems (In the SI Units)
23. Design Concepts in Industrial Hydraulic Systems (In the English Units)
24. Maintenance, Troubleshooting, and Safety in Hydraulic Systems
25. Hydrostatic Transmissions (HSTs) (In the SI Units)
26. Concepts of Hydrostatic Transmissions (In the English Units)
27. Load Sensing Hydraulic Systems (In the SI Units)
28. Concepts of Load Sensing Hydraulic Systems (In the English Units)
29. Electro-hydraulic Proportional Valves
30. Electro-hydraulic Servo Valves
31. Cartridge Valves
32. Electro-hydraulic Systems and Relay Circuits

For more details, please visit: **htpps://jojibooks.com**

About the Author

Joji Parambath is a trainer in the field of Pneumatics, Hydraulics, and PLC, for over 25 years. During his career, he has trained numerous professionals from the industries as well as faculty members and students of engineering institutions.

At present, he is the key trainer at Fluidsys Training Centre, Bangalore, India, (https://fluidsys.org) which is providing training in the field of Pneumatics and Hydraulics. He has already written two books on Pneumatics and Hydraulics. The publication of the present series of 32 books is intended to restructure and update the existing books.

The author wishes to thank all trainees for their lively interaction and many useful suggestions during the training programmes that prompted the author to write the present series of books. You may send your feedback to joji.p@hotmail.com

10th June 2020